Carla Silva
Leonel Nunes
João Matias

Application of an efficiency indicator in the biomass industry

Application of an efficiency indicator in the biomass industry

Carla Silva
Leonel Nunes
João Matias

Application of an efficiency indicator in the biomass industry

Case study on integrating the Lean Manufacturing philosophy

ScienciaScripts

Imprint

Any brand names and product names mentioned in this book are subject to trademark, brand or patent protection and are trademarks or registered trademarks of their respective holders. The use of brand names, product names, common names, trade names, product descriptions etc. even without a particular marking in this work is in no way to be construed to mean that such names may be regarded as unrestricted in respect of trademark and brand protection legislation and could thus be used by anyone.

Cover image: www.ingimage.com

This book is a translation from the original published under ISBN 978-620-2-04751-7.

Publisher:
Sciencia Scripts
is a trademark of
Dodo Books Indian Ocean Ltd. and OmniScriptum S.R.L publishing group

120 High Road, East Finchley, London, N2 9ED, United Kingdom
Str. Armeneasca 28/1, office 1, Chisinau MD-2012, Republic of Moldova, Europe
Managing Directors: Ieva Konstantinova, Victoria Ursu
info@omniscriptum.com

Printed at: see last page
ISBN: 978-620-2-73762-3

Index

keywords   *Lean Manufacturing*, torrefied biomass *pellets*, *Overall Equipment Efficiency* and 5S.

summary   This project took place in a company that produces torrefied biomass *pellets*. The aim was to apply a *lean* methodology in order to improve workplace conditions for the company's employees. This methodology was 5S, to which a new S was added - safety. The production process was also characterized using the *Overall Equipment Efficiency* indicator, before and after the implementation of the 5S+1 methodology. In the end, it was found that there was no improvement in this indicator, and it can be concluded that the methodology used may not have been the most appropriate for the situation under analysis, although there were also some constraints caused by the company's schedule of forced stoppages.

1 Background and Objectives

This chapter provides a framework for this project, as well as its objectives.

This project came about as a result of an internship at a biomass energy company called Yser Green Energy SA. Due to the high production stoppages in this company, the opportunity arose to apply the (5+1) S methodology in order to optimize the production line, which is one of the objectives of this project.

However, this methodology needs to be evaluated, so the other objective will be to characterize production efficiency before and after applying this methodology. This will be done using the *Overall Equipment Efficiency* methodology.

2. Introduction

This chapter provides an introduction to the market associated with the company involved in this project.

Nowadays, there has been a growing demand for comfort, which has led to a huge increase in energy. This increase, mainly in the use of fossil fuels, has consequently led both to an increase in prices, which is a very important factor in competitiveness between companies, and also to a greater release of greenhouse gases and other gases that are harmful to the environment and to people's health due to their combustion (L. J. R. Nunes, Matias, and Catalao 2016).

In order to avoid climate change, there are alternatives, such as the use of biomass (L. J. R. Nunes, Matias, and Catalao 2016; L. J. R. Nunes, Matias, and Catalao 2016), which is a natural and endogenous resource that, as it is produced domestically, is not subject to international price fluctuations (L. J. R. Nunes, Matias, and Catalao 2016). However, due to the low energy density and low yield per unit area of biomass, biomass *pellets* have emerged which have a lower moisture content, a higher energy content and better combustion efficiency (Monteiro, Mantha, and Rouboa 2012).

Pellets can be produced from different raw materials such as pine, eucalyptus, forestry residues, agricultural and forestry residues (Monteiro, Mantha, and Rouboa 2012; L. Nunes, Matias, and Catalao 2014). In Portugal, the most widely used raw material is Pinus pinaster. However, there are regions where other species are used because companies are not interested in obtaining raw materials from a distance of more than 70 km, as this entails high transportation costs (L. J. R. Nunes, Matias, and Catalao 2016).

Pellet combustion equipment is increasingly developed, making maintenance little or easy. This leads to greater interest on the part of the end user (L. J. R. Nunes, Matias, and Catalao 2016). Thus, the *pellet* industry becomes a great opportunity for countries that have forest resources, such as Portugal. If these resources are well exploited and abundant, they can be sustainable (L. J. R. Nunes, Matias, and Catalao 2016).

Pellet production in Portugal is mainly for export, namely to northern Europe (Monteiro, Mantha, and Rouboa 2012; L. J. R. Nunes, Matias, and Catalao 2016; L. Nunes, Matias, and Catalao 2014). However, among national consumers, the main sectors consuming *pellets* are the domestic sector, public services and industries requiring thermal energy (L. J. R. Nunes, Matias, and Catalao 2016).

In addition to biomass pellets, there are torrefied biomass *pellets*. These *pellets* have better properties than traditional *pellets*: higher calorific value, higher bulk density, greater durability, greater resistance to biological activity and are hydrophobic (L. J. R. Nunes, Matias, and Catalao 2016). In addition, they can be used as a substitute for coal, as their properties are similar to those of bituminous coal (L. J. R. Nunes, Matias, and Catalao 2016). However, torrefied biomass *pellets* have higher processing costs due to torrefaction, which high amounts of thermal energy (L. J. R. Nunes, Matias, and Catalao 2016).

Pellet production in Portugal has been on the rise. This increase also leads to the emergence of new companies, which makes this market more competitive. Therefore, in order to be able to offer a more favorable price to the customer, it is necessary to take into account the costs incurred during production, particularly with waste. This leads to the need to use tools to eliminate this waste, which leads to improvements in production and cost reductions. These tools are *lean* methodologies.

3. Characterization of the company

This chapter describes the company where this project is being carried out.

Yser Green Energy, SA (YGE) is a company belonging to the Yser group, whose objective is research and development related to renewable energies from biomass. The target market is the energy market, both industrially and domestically.

YGE focused on the production process of torrefied biomass *pellets*, using biomass such as pine, eucalyptus and acacia. The research unit has the capacity to produce 722 kg/h of torrefied biomass *pellets*, or 5,718 tons/year.

In a nutshell, the production process begins by receiving the raw material, then the raw material is broken down so that it can enter the production line. The raw material is then ground, dried in a rotating drum and ground again. The raw material passes through the roasting reactor, into the roaster, on to the augers in order to humidify and cool it down, then it is *pelletized* and finally the final product is stored, packaged and shipped. This process is shown in Figure 1.

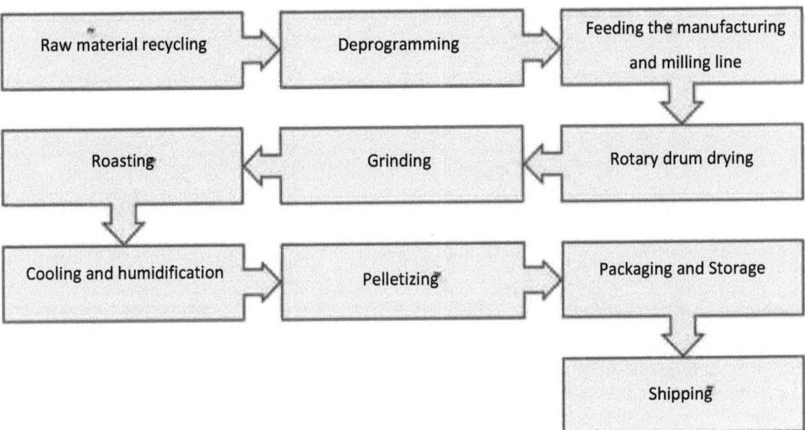

Figure 1: YGE's summary production process.

YGE is not currently producing torrefied biomass *pellets*, but "white" *pellets* i.e. dry biomass *pellets*. It therefore has a higher production capacity of around 1000kg/h.

4. Methodology

This section will present the methodology of this project.

To apply this project, I will characterize the company's production process. I will then begin by calculating the *Overall Equipment Efficiency* of the company's *pelletizer*, as this is the machine that makes the last change to the product.

After this calculation, the *lean* methodology called 5S will be applied, which will be carried out in the order of its five stages. In addition to the 5S relating to this methodology, a new "S" will be added, which refers to safety.

After this application, the OEE will be calculated again to check whether or not it has improved.

5. State of the art

This chapter presents a review of concepts related to the *Lean Manufacturing* philosophy.

5.1. Origin of *Lean Manufacturing*

Toyota Production System

Due to the scarcity of human, financial and material resources after the Second World War, Japan saw the USA dominate the automobile market due to the evolution of Henry Ford's production system. It was then that Toyoda Kiichiro, president of *Toyota Motor Company* (TMC), decided to start studying a new production system, which was called the *Toyota Production System* (TPS) (Almeida 2015; Freitas 2014).

The TPS emerged as a fusion between the Ford production system and the artisanal production system (Almeida 2015) and aimed to remove waste and unconsciousness from the production system in order to maximize production flow (Chowdary and George 2013; Jasti and Kodali 2014).

It is through the House of TPS that this system is usually presented. Figure 2 shows that the pillars of TPS are the *Just-in-Time* and *Jidoka* philosophies, and the basis is level production, visual management and standardized work. In order to deliver value to *stakeholders*, i.e. lower costs, faster response and higher quality, it is necessary, in addition to the pillars and the basis, to work as a team and eliminate waste (B. D. O. Nunes 2014).

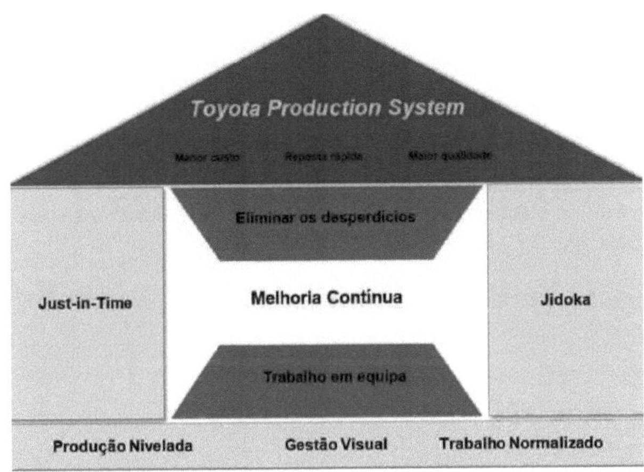

Figure 2: Home of the TPS (Adapted from Nunes, B. D. O., 2014).

Lean Manufacturing

The lean philosophy began with the introduction of TPS (Chowdary and George 2013; Jasti and Kodali 2014). But it was Womarck who popularized the concept of *lean* in 1993 with his book "The Machine that Changed the World" (Cherrafi et al. 2016).

The main objective of *lean* methodologies is to eliminate waste by making the production of products or services of higher quality, in a shorter time and at a lower cost, i.e. to make the process more efficient and streamlined (Cherrafi et al. 2016; Jasti and Kodali 2014; Kull et al. 2014). The application of *lean* methodologies reduces lead times, which improves the delivery of products within the agreed timeframe (Chowdary and George 2013).

Companies believe that by consistently and disciplinedly applying *lean* strategies, it is possible to achieve excellence (Chowdary & George 2013). However, the application of *lean* requires the involvement of people, which can lead to differences in application depending on the culture (Cherrafi et al. 2016; Kull et al. 2014).

1.1. *Lean* principles

The *lean* philosophy is being widely implemented by companies. According to Womack and Jones, the principles of the *lean* philosophy are (Freitas 2014; J. P. M. Gomes 2014; Pinto 2014):

1. **Creating value:** This is the first step in implementing the *lean* philosophy. Value is defined by the customer. Since the characteristics of the product or service that do not please the customer represent opportunities for improvement, the company must continually rethink value in order to create a product that meets the customer's needs;

2. **Defining the value chain:** The value chain is the set of actions that add value to the product or service until it reaches the customer. It is divided into three management activities: development (design, conception and delivery of the product), information management (monitoring orders) and physical transformation (transformation of the product or service). However, it is necessary to manage the chain as a whole. Waste can only be eliminated in activities that generate value for the product or service and that can be linked together in a chain. To understand what these activities are, it is possible to identify them in three types: activities that add value to the product or service, activities that don't add value but are necessary and activities that don't add value and can be eliminated because they are unnecessary;

3. **Optimizing the flow:** Having recognized the value and defined the value chain, it is essential to create a continuous flow of materials, information, people and money. This allows waste to be eliminated and makes the process smoother and more fluid. This process is done in three steps: (1) keeping the focus on the product or service itself from the beginning to the end of the Value Chain, (2) eliminating all obstructions to the continuous flow of that product or service, (3) reflecting the techniques, tools and work practices in such a way as to eliminate possible backtracking, waste and stoppages, to allow the design, customer order and production of a product to succeed continuously;

4. **Pull system:** By creating a continuous flow, the most observable consequence is a reduction in lead-time, i.e. companies are able to respond more quickly to customer requirements. In a pull system, production only starts once the customer has placed an order, thus avoiding overproduction and unnecessary inventory, although inventory is always needed due to production systems that don't allow one product to be produced at a time. This system means that we produce according to customer needs;

5. **Perfection:** After the previous phases, the company is ready to achieve perfection. However, perfection is like infinity, i.e. it is impossible to achieve, which means that companies are constantly reducing effort, errors, space used, delivery time, costs and processing.

These principles are applied in the order in which they were presented and allow waste to be eliminated (J. P. M. Gomes 2014; Pinto 2014).

1.2. Pillars of the *Lean* Philosophy

As mentioned above, the two pillars of the *lean* philosophy are *Just - in - Time* and *Jidoka*.

Just - in - Time

O *Just-in-time* (JIT) is a production system whose aim is to adjust the pace of operations according to the needs of downstream operators, i.e. customers. In this production system, only what is needed, in the quantity needed and what is required is produced in order to ration equipment, capital and labor resources (Silva 2014; Valente 2012).

In order to implement JIT, the *layout* needs to be organized in such a way that there are small production lines, but not too far apart, in order to reduce time between lines and in preparing equipment and machines (Silva 2014; Valente 2012).

Jidoka

The term *Jidoka* means intelligent automation. This is a method of quality control, as it gives the equipment the ability to distinguish parts with defects from parts without defects without the need for an operator to keep an eye on them (J. P. M. Gomes 2014; Gordon 2015).

Once defects have been detected, procedures are triggered to solve the problem, starting with stopping production, followed by the operator correcting the problem (Gordon 2015; Valente 2012).

This process allows defective parts not to continue in the production process, the causes of defects to be easily located, no need for inspection at the end of the process and a reduction in the number of operators in the process (Valente 2012). *Thus, Jidoka* is a prerequisite for *Just-in-time*, because without this system it is not possible to produce what is needed in the time required, due to the existence of defective parts (Gordon 2015).

1.3. Waste

Waste is any activity and resource that does not create value for the product (Dewi, Setiawan, and Susatyo Nugroho 2013; Wahab, Mukhtar, and Sulaiman 2013). Chowdary & George (2013), Dewi et al. (2013) and Cherrafi et al. (2016) state that there are seven forms of waste: *inventory*, transportation, overproduction, movement, waiting, overprocessing and defects. These are the forms that customers don't want to pay for, yet they have a direct impact on quality and price (Cherrafi et al. 2016). In contrast to waste, there are other activities that add value to the product, such as all raw material processing activities (Dewi, Setiawan, and Susatyo Nugroho 2013).

Inventory (*stock*)

Excess raw materials, semi-finished products and finished products come from overproduction, differences in the production capacities of processes, quality problems and early production (B. D. O. Nunes 2014). Excessive stock leads to storage costs, along with transportation and resource costs (A. L. F. Gomes 2014; Silva 2014). To eliminate this stock, it is necessary to produce the exact quantities at the exact time (Valente 2012).

Transportation

Transportation is necessary, but sometimes materials, people or even information are transported unnecessarily (A. L. F. Gomes 2014). This unnecessary transportation can result from a poorly designed *layout* (J. P. M. Gomes 2014).

Overproduction

Producing more products than needed or too soon leads to excessive stock, unnecessary use of resources, excessive consumption of materials and rigid planning (Freitas 2014; J. P. M. Gomes 2014; Valente 2012). This is due to the need to reduce machine set-up times (A. L. F. Gomes 2014).

Movement

Excessive movements of operators and machines are often not considered to be processes that do not create value, as they are integrated into other processes (J. P. M. Gomes 2014; Silva 2014). These movements can arise due to the lack of training and/or motivation of operators, the instability of operations, lack of storage of space and tools and inappropriate *layout* (A. L. F. Gomes 2014; J. P. M. Gomes 2014; Valente 2012).

Waiting

Waiting time refers to the time the operator or equipment is idle (Almeida 2015; A. L. F. Gomes 2014). This happens because information, equipment, materials or people are not available, due to lack of work, breakdowns, lack of raw materials, tool changes, among others (Almeida 2015; Silva 2014; Valente 2012). Waiting times lead to delays in delivery times (A. L. F. Gomes 2014; Valente 2012).

Excessive processing

Over-processing refers to processes that do not add value to the product (Silva 2014). This refers to processes for producing products of a higher quality than the customer wants or needs, retouching and repair processes, incorrect handling of tools and equipment and complex or incorrect procedures (A. L. F. Gomes 2014; Silva 2014; Valente 2012).

Defects

Defects are parts that are produced outside the predefined quality specifications (A. L. F. Gomes 2014; Silva 2014). This is a very common defect in industry and causes losses, including monetary losses due to raw materials, labor, machinery, handling, among others (Cruz 2013; Valente 2012).

1.4. *Lean* tools

Overall Equipment Efficiency (OEE)

Overall Equipment Efficiency (OEE) is increasingly being used in industry, as are *lean* methodologies in general (Puvanasvaran, Kim, and Siang 2012).

OEE measures the effectiveness of a piece of equipment in relation to its potential potential and takes into account availability, equipment performance, efficiency losses and performance losses (Azizi 2015; Puvanasvaran, Kim, and Siang 2012).

OEE emerged in the *Total Productive Maintenance* (TPM) approach, which is part of TPS (Almeida 2015). As companies want to have their equipment working at maximum efficiency, by analyzing the OEE and its individual factors it is possible to see where the most time is being lost in relation to its potential (Puvanasvaran, Kim, and Siang 2012).

The large losses in the different factors of the OEE calculation come from (Almeida 2015; Azizi 2015; Puvanasvaran, Kim, and Siang 2012):

• Breakdowns: these can be mechanical, electrical, lack of tools, general equipment failure or stops that were not planned for maintenance and lead to a stoppage of more than 5 minutes. This loss leads to a reduction in the time the equipment is available for production;

- Set-up and adjustments: these reduce the time available for the equipment in question and are due to tool replacements, material shortages, production changes and cleaning stops;

- Minor stoppages: stoppages that do not exceed 5 to 10 minutes and are caused by breaks in the production line, which can be due to blockages in the production flow, cleaning or minor adjustments:

- Reduced speed: production at a lower speed than that specified for the equipment leads to a difference between the quantity produced and the theoretical quantity, i.e. the quantity that could be produced;

- Rejection at start-up: during the start-up phase, the equipment can produce defective products, due to the time it takes to restore technical limitations, for example, preheating and tuning errors.

- Defective production: reduces the amount of product that meets specifications and can be caused by equipment malfunction or operator misoperation.

However, there are also planned stoppages. Some companies prefer to include them in the OEE calculation in order to reduce these stoppages, which can be due to lack of orders, meetings or tests (Almeida 2015).

The OEE depends on the availability of the equipment, the performance rate, i.e. efficiency, and the quality of the product and is calculated by multiplying these three parameters (Azizi 2015; Puvanasvaran, Kim, and Siang 2012):

$$OEE = Availability \cdot Efficiency \cdot Quality \tag{1}$$

Availability is calculated taking into account the actual production time, i.e. the time during which production was actually carried out, and the planned production time, i.e. the time during which production should have been carried out (J. P. M. Gomes 2014; Valente 2012):

$$Availability = \frac{Time\ to\ Produce\ the\ Real}{Planned\ Production\ Time} \tag{2}$$

To calculate production time, you need to know the shift hours and subtract unplanned downtime. The planned production time is calculated by subtracting the number of shift hours from the number of hours of planned downtime.

With regard to efficiency, you need to know production data, namely current production and expected production. Current production is the quantity or speed produced today or over the course of data collection, and expected production is the quantity or speed that could be expected to be produced, taking into account the equipment's capabilities. This calculation is made by dividing current production by expected production (Almeida 2015):

$$Efficiency = \frac{Produce\ the\ Real}{Expected\ Output \cdot Planned} \tag{3}$$

$$Quality = \frac{Quantity\ of\ non\text{-}defective\ product.}{Total\ Product\ Quantity} \tag{3}$$

Quality is given by dividing the number of non-defective products, i.e. products that meet specifications, by the total number of products (Almeida 2015; J. P. M. Gomes 2014; Valente 2012):

(4)

The OEE requires precise and careful analysis in order to decide on improvement actions. These improvement actions should focus mainly on the points where faster results can be obtained (Almeida 2015; Valente 2012).

Overall Factory Effectiveness (OFE)

Coтo refendo, OEE is only calculated for one piece of equipment. However, the important thing is to improve not the performance of one piece of equipment, but that of a factory. To address this, the concept of *Overall Factory Effectiveness* (OFE) emerged. This concept is still under development and allows for the combination of activities, interactions between equipment and processes and the integration of information, actions and decisions in independent systems (Oechsner etal. 2003).

In order to evaluate the OFE, it is necessary to take into account the OEE, material costs and energy efficiency, efficiency, yield, performance, etc. A good way of understanding how to improve OEE and OFE is through simulation (Oechsner et al. 2003).

(5 + l)S

The 5S tool emerged in Japan due to the application of the *Kaizen* methodology (Jiménez et al. 2015). As the name suggests, this tool consists of five (Japanese) words beginning with "s". These words are steps to follow when applying this tool in order to improve working conditions to obtain and/or maintain a quality environment (Dewi, Setiawan, and Susatyo Nugroho 2013; Jiménez et al. 2015; Sahu, Patidar, and Soni 2015). However, in addition to implementing the Ss, it is necessary to change people's thinking, behavior and attitude (Neves 2015).

The application of this tool brings several advantages, among them (Freitas 2014; Silva 2014): improved working conditions, reduced costs, reduced waste, increased quality, increased safety, among others.

According to (Freitas 2014; Jiménez et al. 2015; Neves 2015) the different phases of the implementation of this tool are:

- *Seiri* (Organize) - Removing unnecessary[i] items from the workplace. This phase requires common sense in order to classify objects and documents so as to define which are used and which are not by users. After this classification, it is necessary to define what can be kept at the workplace, what can be stored elsewhere and what can be discarded;

- *Seiton* - Organizing materials so that they can be used more easily and returned to their place after use. It is necessary to define storage locations and standards, and it is also interesting to implement visual systems;

- *Seiso* (Clean) - Maintaining the cleanliness of the workplace. The aim is to keep the space clean at all times and to change any actions that could make the space dirty. Cleaning schedules and checklists can be defined;

- *Seiketsu* (Standardize) - Promoting the standardization of the operations in the previous points. People should be encouraged to maintain personal and workplace hygiene in order to keep the place in good condition at all times and ensure that the environment is always kept clean and tidy. This allows for the control and maintenance of the previous 3S;

- *Shitsuke* (Self-discipline) - Teaching the 5s practices so that they are used continuously. This phase aims to correct inappropriate behavior and shape habits in order to maintain the efforts made in the previous 4S. This phase is more complicated due to the fear and complacency associated with human beings when faced with change.

---

[i] An item is any material, tool, spare part, document or information.

In order for all employees to be sensitive to safety issues in the workplace, there is the possibility of extending the application of the 5S methodology to one more, which would be related to safety.

This new S is intended to prevent potential accidents and eliminate hazards and cannot be disaggregated from the other S's (Citeve 2012; Li et al. 2016).

To help with the implementation and continued use of 5S, it is important to use visual aids.

6. Characterization of the production process and application of *Lean Manufacturing*

This chapter will present the characterization of the production process. In addition, the OEE indicator will be calculated, followed by the implementation of the 5S methodology. Finally, the OEE indicator will be calculated again.

6.1. Characterization of the production process

YGE's production process begins when the raw material is received. This raw material (Figure 3), as mentioned above, can be pine, eucalyptus or acacia, depending on the availability and price of the raw material.

Figure 3: Raw materials at YGE.

To enter the production process, the raw material is placed in a shredder (Figure 4) in order to obtain the shredded raw material ready for the production process (Figure 5).

Figure 4: YGE shredder.

Figure 5: Wrecked raw materials.

The raw material is then placed in the *Self Feeder* (Figure 6) so that it can enter the production line under control. After entering the production line, the raw material passes through a hammer mill (Figure 7) which reduces it in size.

Figure 6: YGE *Self Feeder*.

Figure 7: *Self-Feeder* and Hammer Mill.

The raw material then passes through a cyclone and is dried in the dryer (Figure 8). After drying, the raw material passes through a double cyclone and is ground again in a hammer mill (Figure 9).

Figure 8: Cyclone and dryer in YGE's production process.

Figure 9: Double cyclone and second hammer mill in YGE's production process.

After this process, the raw material goes through a sieve in order to separate the fine particles from the coarser ones, with the finer ones, if necessary, going to the fuel tank in order to be burned in the furnaces (Figure 10) that heat the air in the dryer and the roasting reactor.

Figure 10: YGE *furnace*.

The raw material then passes through the roasting reactor (Figure 11), which has been out of operation; it is currently used to store the raw material as no roasted *pellets* have been produced. Afterwards, the raw material goes to a mixer (Figure 12), through augers that have a water cooling system.

Figure 11: Roasting reactor.

Figure 12: Mixer.

This is followed by pelletizing (Figure 13), where it is necessary to control the amount of water added to the raw material in order to pelletize as much raw material as possible.

After pelletizing, the *pellets* go to a sieve, where they are separated from particles that have not been pelletized. The *pellets* then go to the storage silos (Figure 14) and the remaining particles return to the mixer to be pelletized later.

Figure 13: Pelletizer.

Figure 14: Silos for storing pellets.

6.2. Initial analysis of the OEE indicator

OEE is only calculated for one piece of equipment, but in this case, it is calculated for the entire production line, because if a problem occurs at any point on the production line, the entire line comes to a halt. Therefore, the piece of equipment selected for calculating this indicator was the pelletizer, as this is the last piece of equipment to be transformed into the final product and is the most important piece of equipment in production.

The OEE was calculated based on what is described in Section 3.5 and taking into account production data from eight working days - from January 16th to 20th and 23rd to 25th. The production data, namely actual production time, planned production time, actual production and planned production in terms of quantities, can be found in Table 1.

It should be noted that the values relating to the quantities of defective product and total products are not broken down in the table, as they are already implicit in the value of actual production, since this only accounts for the product that meets the customer's requirements.

Table 1: Production data for the days under analysis for OEE.

Day	Actual Production Time(h)	Planned Production Time (h)	Actual Production (ton/day)	Planned Production (ton/day)
1	10,10	24,00	1,98	24,00
2	22,90	24,00	16,75	24,00
3	18,02	24,00	15,54	24,00
4	23,23	24,00	22,38	24,00
5	8,90	9,00	8,26	9,00
6	16,21	16,00	16,145	16,00
7	14,70	16,00	14,43	16,00
8	12,42	24,00	11,00	24,00
Sum	126,48	161,00	106,485	161,00

Based on the data in the table above, it is possible to calculate both availability and production efficiency:

$$Disponibilidade = \frac{Tempo\ de\ Produ\_o\ Real}{Tempo\ de\ Produ\_o\ Planeado} = \frac{126,48}{161,00} = 0,785 \implies \boxed{78,5\%}$$

$$Efici\_ncia = \frac{Produ\_o\ Real}{Produ\_o\ Esperada} = \frac{106,485}{161,00} = 0,661 \implies \boxed{66,1\%}$$

After calculating the availability and efficiency of the equipment, i.e. the production line, it is possible to calculate the OEE value.

$$OEE = Availability \cdot Efficiency \cdot Quality = 0.785 \cdot 0.661 = 0.519 \implies \boxed{51,9\%}$$

Looking at the OEE value (51.9%), it can be seen that it is low and needs to be improved. The 5S methodology will therefore be applied.

6.3. Implementation of {5+I) S

Initial state of the company

Before the implementation of the (5+1) S methodology, it was possible to see that the workplace of the employees in the production area was not in the best condition. Figure 15 and Figure 16 show the lack of organization and storage of maintenance tools in the production area. Figure 17 and Figure 18 show that the tanks in the production area and the production control room are not clean. It should be noted that this is a company whose production line emits a lot of dust, which makes it tolerable that some places are not always clean.

Figure 19 shows the lack of safety when placing pallets near a room in the production area that can be frequented by forklifts.

Figure 15: Initial state of the company in terms of organization and tidiness.

Figure 16: Initial state of the company in terms of organization and tidiness.

Figure 17: Initial state of the company in relation to cleanliness.

Figure 18: Initial state of the company in relation to cleanliness.

Figure 19: Lack of security in the company.

Implementation of the methodology (5+1) S

The implementation of the (5+1)S methodology requires the cooperation of the company's employees, as they are the main stakeholders and beneficiaries of this implementation. Therefore, before applying the (5+1)S methodology, employees were made aware of the methodology. This awareness-raising facilitates employee participation in the application of this methodology.

As mentioned in the State of the Art, this methodology was followed in the order of the Ss. Table 2 shows the points made at each stage. These points are described later.

Table 2: Points scored in each stage of the (5+1) S methodology.

Stage	Points to be achieved
1ᵉS	Separate the useless items from the useful ones.
	Keep all useful items on site.
	Eliminate unnecessary items.
	Eliminate unnecessary information.
2ᵉS	Identify storage areas and safety areas at workstations.
	Identify the items and places to store them in the workplace.
	Arrange each item in its place so that the most used items have easier access.
3ᵉS	Cleaning the workplace, including items, storage areas and safety areas
	Keep cleaning materials in the designated place.
4ᵉS	Drawing up organizational rules and cleaning rules.
	Disseminate the rules of organization and cleanliness.
	Make employees, particularly those responsible for compliance with the rules, aware of the importance of compliance.
5ᵉS	Make the person responsible for complying with the rules of organization and cleanliness aware of the importance of carrying out regular audits.
6ᵉS	Draw up a list of safety rules.
	Make those responsible for compliance with safety rules aware of the importance of carrying out regular audits.

1- <u>Step: Se/r/-organize</u>

At this stage, a survey was carried out of the items present in each workplace implementing the (5+1)S methodology. Subsequently, and with the cooperation of the employees, as mentioned in the State of the Art, the usage rate of each item was assessed in order to keep the useful items in the workplace and the useless ones eliminated or placed elsewhere. It should be noted that Items include material and information.

<u>2ᵍ EtapaiSezton-Arrumar</u>

At this stage, the storage locations for each item were defined, depending on their use. The items were then arranged, as can be seen in Figure 20 and Figure 21.

Figure 20: Storage of maintenance material.

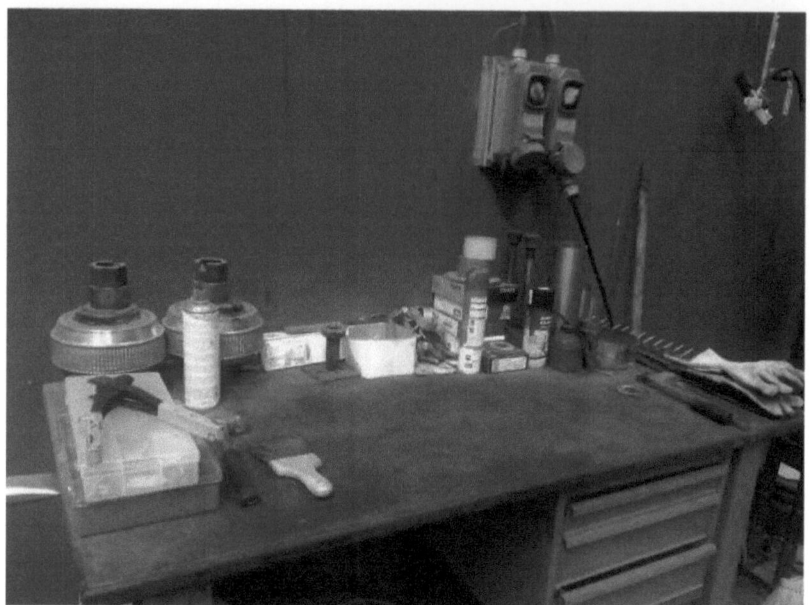

Figure 21: Arrangement of the maintenance table with some tools.

<u>3ª Step: *Seiso* - Clean</u>

This phase was carried out in parallel with the previous stages, so that the workplace would already be clean when the items were put away.

It was necessary to define the cleaning periods for each area, so it was essential to make the workers aware of the need for regular cleaning and it was decided that the shift manager would be responsible for checking the cleanliness at the established periods.

Figure 22 and Figure 23 show how clean the factory is.

Figure 22: Iimpa production area.

Figure 23: Secretariat of the Iimpa Control room.

4- Stage: *Seiketsu* - Normalize

At this stage, the organization and cleaning rules were defined, which can be found in Annex A and Annex B, respectively.

And the rules defined earlier were standardized and formalized so that employees adopt these new rules.

The aim of this phase is to ensure that employees don't slip back into their previous habits. This will be done by the shift manager.

5ᵉ Stage: *Shitsuke* - Self-discipline

At this stage, it is necessary to ensure that the rules are complied with, and to this end, compliance must be monitored. This monitoring needs to be more regular in the beginning, because changing habits is difficult so that employees don't regress to old habits. This leads to positive results after the implementation of 5S.

The audit of compliance with these rules will be carried out by the shift manager.

6- Stage: Security

This phase was carried out at the same time as the others, as the items should be in places that do not interfere negatively with employee safety. Employees were made aware of the importance of maintaining safety equipment (fire extinguishers and emergency showers and eyewashes), as well as the importance of complying with the safety rules, which can be found in Appendix C.

Safety rules will be audited periodically or always, depending on the rules, by the shift manager.

Figure 24 shows the signposting of some safety rules.

Figure 24: Safety rules signage.

1.   4. Analysis of the OEE indicator after implementation of the methodology (5+1) S

As with the initial calculation of the OEE indicator, the assessment is made for the entire production line, because if a problem occurs at any point on the production line, the entire line is stopped. Therefore, the equipment selected for calculating this indicator was the *pelletizer*, as this is the last piece of equipment to be transformed into the final product and is the most important piece of equipment in production.

The OEE was calculated based on what is described in Section 3.5 and taking into account production data from nine working days - from May 2 to 5 and from May 8 to 12. The production data, namely actual production time, planned production time, actual production and planned production in terms of quantities, can be found in Table 3.

As before, it should be noted that the values relating to the quantities of defective product and total products are not broken down in the table, as they are implicit in the value of actual production, since this only accounts for the product that meets the customer's requirements.

Table 3: Production data for the days under analysis for OEE.

Day	Actual Production Time(h)	Planned Production Time (h)	Actual Production (tons per day)	Planned Production (ton/day)
1	13,42	16,00	12,36	16,00
2	12,20	16,00	9,96	16,00
3	8,52	16,00	4,77	16,00
4	6,61	16,00	2,70	16,00
5	0,00	8,00	0,10	8,00
6	13,48	16,00	11,62	16,00
7	14,82	16,00	14,22	16,00
8	8,68	16,00	7,36	16,00
9	0,00	8,00	0,92	8,00
TOTAL	77,73	128,00	64,01	128,00

Based on the data in the table above, it is possible to calculate both availability and production efficiency:

$$Disponibilidade = \frac{Tempo\ de\ Produ\_o\ Real}{Tempo\ de\ Produ\_o\ Planeado} = \frac{77,73}{128,00} = 0,607 \implies \boxed{60,7\%}$$

$$Efici\_ncia = \frac{Produ\_o\ Real}{Produ\_o\ Planeada} = \frac{64,01}{128,00} = 0,500 \implies \boxed{50,0\%}$$

After calculating the availability and efficiency of the equipment, i.e. the production line, it is possible to calculate the OEE value.

$$OEE = Disponibilidade Efici\_ncia Qualidade = 0,607 0,500 = 0,304 \implies \boxed{30,4\%}$$

If you look at the OEE value (30.4%), you can see that it is lower than the OEE value before the implementation of the (5+1) S methodology, which was 51.9%.

7. Discussion of results

This chapter briefly discusses the results obtained in relation to the OEE indicator.

As mentioned above, the final OEE value is lower than the initial value. This can be due to a number of reasons. Firstly, despite the implementation of the 5S methodology, production stoppages continue to come from the same source, including problems on the production line, such as jamming at various points on the production line, such as hammer mills, cup conveyors, pelletizing machines, among others. Another reason for production stoppages is a lack of raw materials.

It should be noted that the 5S methodology is not the most suitable for improving the OEE indicator, since its main objective is to improve workplace conditions, such as organization and cleanliness. While the OEE indicator is more related to maintenance, a methodology that could be implemented in the future in order to increase the value of the indicator is the *Total Productive Maintenance* methodology.

8. Conclusion

This chapter presents the main conclusions of this project.

Lean Manufacturing helps companies to evolve in line with market demands, as it allows companies to eliminate and reduce waste. For this reason, many companies are implementing *lean* methodologies.

The objectives of this work were to implement the 5S methodology with the addition of a new S relating to safety, to characterize the production line by evaluating the OEE indicator and, finally, to publish a scientific article or conference paper. The first two objectives have been completed, but the last one, due to lack of time, has not been completed to date.

The Overall Equipment Efficiency indicator was evaluated before and after the implementation of the *lean* 5S methodology. It would have been expected that the value of the OEE indicator would increase, but this was not the case, because in order to improve OEE it was necessary to improve each index individually (availability, efficiency and quality), which was not done in this project.

This situation was also due to the company's own constraints, since it is also a research and development unit, and it was necessary to interrupt continuous production several times in order to carry out tests with different forms of biomass.

However, despite the fact that the OEE indicator has decreased, the general opinion of the company's employees is that it has become much easier to operate.

The 5S methodology is very adaptable, so it can be applied in any situation or area.

In addition to suggesting the application of the TPM methodology in the Results Discussion, I also suggest the implementation of the visual management tool so that employees can more easily locate the items they need.

9. Bibliography

Almeida, Joao Pedro Tavares. 2015. *Application of Lean Methodologies and Tools in the Production Process of Heliflex, Tubos E Mangueiras, S.A.*

Azizi, Amir. 2015. "Evaluation Improvement of Production Productivity Performance Using Statistical Process Control, Overall Equipment Efficiency, and Autonomous Maintenance." *Procedia Manufacturing* 2(February): 186-90.

Cherrafi, Anass et al. 2016. "The Integration of Lean Manufacturing, Six Sigma and Sustainability: A Literature Review and Future Research Directions for Developing a Specific Model." *Journal of Cleaner Production* 139: 828-46.

Chowdary, Boppana V., and Damian George. 2013. "Improvement of Manufacturing Operations at a Pharmaceutical Company: A Lean Manufacturing Approach." *Journal of Manufacturing Technology Management* 23: 56-75.

Citeve. 2012. "Tool for the Development and Application of Lean Thinking in STV."

Cruz, Nuno Miguel Pereira da. 2013. *Implementation of Lean Manufacturing Tools in the Plastic Injection Process.*

Dewi, Septika Rosiana, Budi Setiawan, and W P Susatyo Nugroho. 2013. "5S Program to Reduce Change-over Time on Forming Department (Case Study on CV Piranti Works Temanggung)." *IOP Conference Series: Materials Science and Engineering* 46: 1-12.

Freitas, Filipa Alexandra Alves de. 2014. *Implementation of Lean Tools in the Madeira Beer Company.*

Gomes, Ana Lucia Figueiredo. 2014. *Lean Improvement Actions in a Production Cell.*

Gomes, Joao Pedro Moreira. 2014. *Application of the Lean Manufacturing Philosophy in an Urban Solid Waste Sorting Center.*

Gordon, James. 2015. "What Is Jidoka and How Can Jidoka Improve Your Business?" *The Leadership Network.*

Jasti, Naga Vamsi Krishna, and Rambabu Kodali. 2014. "Lean Production: Literature Review and Trends." *InternationalJournal ofProduction Research* 53(August): 1-19.

Jiménez, Mariano, Luis Romero, Manuel Dominguez, and Maria del Mar Espinosa. 2015. "5S Methodology Implementation in the Laboratories of an Industrial Engineering University School." *SafetyScience* 78: 163-72.

Kull, Thomas J., Tingting Yan, Zhongzhi Liu, and John G. Wacker. 2014. "The Moderation of Lean Manufacturing Effectiveness by Dimensions of National Culture: Testing Practice-Culture Congruence Hypotheses." *InternationalJournal ofProduction Economics* 153: 1-12.

Li, Shuquan, Xiuyu Wu, Yuan Zhou, and Xin Liu. 2016. "A Study on the Evaluation of Implementation Level of Lean Construction in Two Chinese Firms." *Renewable and Sustainable Energy Reviews* (December): 0-1.

Monteiro, Eliseu, Vishveshwar Mantha, and Abel Rouboa. 2012. "Portuguese Pellets Market: Analysis of the Production and Utilization Constrains." *EnergyPolicy* 42: 129-35.

Neves, Claudinete Simone Cabral. 2015. *Implementation of the 5S Methodology at USF CelaSadde.*

Nunes, Bruno Daniel Oliveira. 2014. *Application of Lean Tools in a Production Cell.*

Nunes, L. J R, J. C O Matias, and J. P S Catalao. 2016. "Torrefied Biomass Pellets: An Alternative Fuel for Coal Power

Plants." *International Conference on the European Energy Market, EEM* 2016- July.

Nunes, L. J R, J. C O Matias, and J. P S Catalao. 2016. "Wood Pellets as a Sustainable Energy Alternative in Portugal." *Renewable Energy* 85: 1011-16.

Nunes, LJR, JCO Matias, and JPS Catalao. 2014. "Economic and Sustainability Comparative Study of Wood Pellets Production in Portugal, Germany and Sweden." *Proceedings of the International* ... 1(12): 526-31.

Oechsner, Richard et al. 2003. "From Overall Equipment Efficiency (OEE) to Overall Fab Effectiveness (OFE)." 5: 333-39.

Pinto, Joao Paulo. 2014. "Introduction to Lean Thinking." *Lean thinking - The philosophy of winning organizations.*

Puvanasvaran, Perumal, Chan Yun Kim, and Teoh Yong Siang. 2012. "Overall Equipment Efficiency (Oee) Improvement Through Integrating Quality Tool : Case Study." (October): 15-16.

Sahu, Shekhar, Lakhan Patidar, and Pradeep Kumar Soni. 2015. "5S Transfusion To Overall Equipment Effectiveness (Oee) for Enhancing Manufacturing Productivity." *International Research Journal of Engineering and Technology* : 2395-56.

Silva, André Castelo Madail da. 2014. *Costing Systems According to Lean Methodologies: Case Study.*

Valente, Filipe Daniel Silva. 2012. *Improving Equipment Availability to Increase OEE.*

Wahab, Amelia Natasya Abdul, Muriati Mukhtar, and Riza Sulaiman. 2013. "A Conceptual Model of Lean Manufacturing Dimensions." *Procedia Technology* 11(Iceei): 1292-98.

Annexes

Annex A: YGE organization rules

Rules of Organization

After use, place each item in its own storage area.

Keep each item in its defined place.

Place the tools on the tool trolley.

Keep the tool trolley in the maintenance area.

Keep the forklift close to the battery charger.

Keep the gas tank in the defined location.

Keep lubricants in the place defined in the maintenance area.

rules

Materials and tools must be clean after use.

The cleaning material must be cleaned after use.

At the end of the first shift each day, clear the production area.

At the end of the second shift each day, clean the production control room.

At the end of the first shift each day, clean the safety equipment.

rules

Always wear protective boots.

Always wear a safety helmet.

Always wear protective earphones.

Wear a protective mask when necessary.

Check safety equipment (fire extinguishers and emergency showers and eyewashes) on the first working day of each month.

yes
I want morebooks!

Buy your books fast and straightforward online - at one of world's fastest growing online book stores! Environmentally sound due to Print-on-Demand technologies.

Buy your books online at
www.morebooks.shop

Kaufen Sie Ihre Bücher schnell und unkompliziert online – auf einer der am schnellsten wachsenden Buchhandelsplattformen weltweit! Dank Print-On-Demand umwelt- und ressourcenschonend produziert.

Bücher schneller online kaufen
www.morebooks.shop

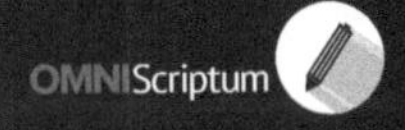